Anthony Quinton Keasbey

Home Flowers pressed in my Law-Books

Anthony Quinton Keasbey

Home Flowers pressed in my Law-Books

ISBN/EAN: 9783743345133

Manufactured in Europe, USA, Canada, Australia, Japa

Cover: Foto ©berggeist007 / pixelio.de

Manufactured and distributed by brebook publishing software
(www.brebook.com)

Anthony Quinton Keasbey

Home Flowers pressed in my Law-Books

HOME FLOWERS

PRESSED IN MY LAW-BOOKS.

COLLECTED CHRISTMAS, 1879,

FOR SUE, BY "Q."

HOLBROOK, PRINT, NEWARK.   NOT PUBLISHED.

CONTENTS.

CONTENTS.

TO SUE.

In years long gone my vow was made
 To send each Christmas day, dear Sue,
Some book whose votive page should bear,
 For friendship's token, " Sue from Q."

And there they stand in lengthening line,
 Sure chroniclers of fleeting years ;
We count them not—'t were vain to sum
 The measure of life's smiles and tears.

Let stranger hands their pages turn
 For all the varied lore they teach,
To us far other voices come
 From those " dumb mouths," in silent speech.

We hear the sounds of love-bound homes,—
 The elders' joy,—the children's glee,—
The lovers' vows,—the marriage bells,—
 The birth-day romps,—the Christmas Tree.

We hear once more from lips long sealed,
   The prayers of love, the counsels wise,
So gently breathed by saint and sage
   Now havened safe beyond the skies.

We hear—alas that words so sad
   My voiceless messengers should tell !
The yearning cry for treasures lost,
   The wail of grief,—the funeral knell !

And now one offering more I make ;
   One more mute mouth whose voice shall blend
With tones that thrill through vanished years
   And reach our hearts alone,—I send.

Not burning words in lofty rhyme,
   Nor wisdom shrined in curious tome,—
A sheaf of wayside flowers that grew
   Untrained beside the porch of home.

Its voice no stranger ears must hear,
   It bears no message save to you,
But whispers to your heart alone
   The steadfast love of Q. for Sue.

1879.

SPRING.

The Church.

When first in ancient days the Eastern star
    O'er Asia's plains its mystic radiance flung,
A corner-stone was laid, and fairer far
    Than earth's proud monuments, the fabric sprung.
The Christian standard from its towers unfurled
Waved wide its sign of mercy o'er the world.

Its glory grew with every battle shock,
    In vain from age to age its foes have striven,
'Twas founded firm on truth's eternal rock ;
    'Twas built by Him who framed the walls of heaven
Its solemn aisles with heavenly music rang,
For angel choirs its earliest anthem sang.

Its course is onward— e'en the ocean isles
    Where night has brooded, hail the wakening dawn ;
The wild is blooming and the desert smiles,
    The reign of darkness and despair is gone ;
O'er India's plains the living seeds are sown,
The Book of Life has pressed the idol's throne.

It shall be onward—spread its portals wide,
   And welcome all from earth's remotest clime,
Till all her nations pour their joyous tide
   To fill its ample courts in coming time ;
Till up from earth's wide altar to the skies,
   The incense of a ransomed world shall rise.

1842.

The Star of the Nativity.

NEW light gleamed o'er the Eastern sands,
   A new star set on high ;
The herald-star of mercy's dawn
   Flamed in the morning sky.
The sages saw, and glorious hope
   To their glad hearts was given,
For with those new-sped beams came down
   A voice of joy from Heaven.

It led them by its guiding ray,
  As o'er the sands they trod,
And like the sun on Gibeon's vale
  Stood o'er their infant God ;
They bowed their reverend heads in awe
  Before that Child Divine,
And proudly poured their treasured wealth
  Upon that lowly shrine.

And where is now that guiding star,
  That watch-fire of the skies,
That burned of old above the shrine
  Of Heaven's great sacrifice ?
When night at noon on Calvary fell,
  And vail and rocks were riven,
Returned it to its home of light
  Beyond the loftiest heaven ?

Not so ! the light the wise men saw
  Was caught by every star,
And still its guiding beam doth fall
  From each bright orb afar.
Though veiled by night with darkening clouds,
  Though lost in day's broad glare,

It points the way for willing feet
　　To every house of prayer.

On every shrine where human hearts
　　For praise or worship bow.
On lowly church or lofty fane
　　That light is shining now;
And Bethlehem's star its silent watch
　　O'er hallowed ground shall keep.
Till down through all the spangled skies
　　The endless night shall sweep.

Oh then, within the temple gates
　　Be thy glad footsteps borne.
As sages sought that humble shrine
　　On earth's first Christmas morn!
Before the star-crowned altar bend
　　Like them in grateful prayer,
Thy richest wealth of worship bring.
　　Thy risen God is there!

1843.

Italian Vespers.

HARK ! a voice of music stealing
  Through the lofty arches dim ;
Murmuring now, now louder pealing—
  'Tis the Vesper Hymn.

Round the altar bending lowly,
  Vestals bow in saintly guise ;
Chanted prayers, in accents holy,
  Like rich incense rise.

Sweetly o'er the moonlit waters
  Float the lute's low tones along,
Sweetly sing the dark-eyed daughters
  Of that land of song.

Softly breathe the broken hearted,
  Bending o'er the loved one's bier,
Requiems for the soul departed
  To a happier sphere.

Sweeter than that song of gladness,
  Purer than the lute's low tone,
Softer than the song of sadness
  From those mourners lone,—

Rise those vesper hymns to heaven,
  Earth's most grateful harmonies
Wafted by the breath of even,
  To the listening skies.

For from that pure shrine ascending
  To the glorious courts above,
There with heavenly anthems blending,
  Swell those tones of love.

1843.

A Ring found on a Grave,

MARKED, "F. G., 1783"

I come from the dark mansion of decay
  Where rest the dead beneath the weeping willow;
Sad nature's harp on which the cold winds play,
  Her solemn dirges o'er their lonely pillow.
I parted from a sorrowing maiden's finger,
A pledge upon the giver's grave to linger.

Oft have I seen the silent funeral train
  Surround the tomb, a band of broken-hearted,
And leave their burden in death's dark domain,
  While memory lingered with the loved departed.
Few human forms are left in those cold halls
O'er whom no tear of love and sorrow falls.

Youth shorn of all its beauty and its bloom,
  And manhood robbed of all its pride and gladness,
And worn-out age, came crowding to the tomb,
  But round each bier I heard the sounds of sadness.
I saw love strew each new-made grave with flowers,
And memory come to mourn in after hours.

———

But I'm free, I am free from the land of graves;
  No more will I dwell with the dead,
No more will I lie where the yew-tree waves
  O'er my cold and lonely bed.

Again will I glitter on beauty's hand,
  And shine in the merry dance,
Again will I dart through a joyous band
  The lightning of my glance.

Again with fond hearts will I gaily rove
    When the moon keeps watch on high,
And pass as a token of plighted love
    As pure and as endless as I.

Then joy to the rare old graveyard ring !
    From the finger of death I am free ;
And sorrow and mourning away I'll fling,
    And a gay old ring will I be.

1843.

Autobiography of a Simoom.

WITH my banner of cloud unfurled I come
    O'er the sea and the land in wrath ;
Unlocked from the caves of my Arctic home
    I rushed on my desolate path.

O'er the snowy peaks of the Northern hills
   I whirled in my cloudy car :
And the tropic vales and the laughing rills
   I chilled with my frown afar.

To the world of waters I sailed away
   As it lay in its evening rest,
And I swept the glories of dying day
   With a glance, from its glowing breast.

A stately ship in her path of pride
   Came quietly gliding by ;
With full-spread sails and a favoring tide
   She smiled at the frowning sky.

But I swooped in wrath from my far-off home
   On my wings of fire and cloud,—
From the snowy threads of the wild sea-foam
   I wove her a lordly shroud.

Away through the vine-clad hills of France,
   And through many a blooming plain,
O'er the sunny meadows where maidens dance,
   I swept with my fearful train.

I tore the vines from the smiling vale—
  I scattered them on the blast,
And through fields of grain with my scythe of hail,
  Like a mighty reaper passed.

1 fly to the Desert, away, away !
  And see ! o'er the burning sands
A caravan comes with its long array
  And the wealth of the Eastern lands.

Their flaunting train o'er the dreary plain
  On the wings of the wind I'll fling,—
But in fear they bow,—I will spare them now,
  For they worship the desert king !

Now I never will roam from the Simoom's home,
  I'll dwell on this burning plain ;
With my sceptre of sand, in this lonely land
  Dread monarch henceforth I'll reign.
1843.

The faded Beauty to her Mirror.

FRIEND of my girlhood, in whose burnished face
   So oft I've gazed in rapture on my own,
On whose bright page I early loved to trace
   The lines of grace that there all radiant shone !

Say, hast thou lost of late thy magic power
   To give true records to the reader's eye ?
Bear'st thou false witness ?   In youth's waning hour
   Must thine, like all earth's friendships, fade and die ?

No, thou wast ever true ; at rosy morn
   The yawn, the night-cap and the robe-de-nuit,
The half-sealed eyes, the papered hair forlorn,
   Received their hideous images from thee.

And oft, on final glance in evening's gloom,
   The glittering ball-dress, and the jeweled hair,
The studied smile, the cheek of borrowed bloom,
   Came beaming back in mellowed lustre there.

Too true, alas ! for as life's morning fair
    Its beauty and its brightness found in thee,
So now the eye, the cheek, the tortured hair
    Come dimmed and faded sadly back to me.

Awful reflection ! must I then resign
    The laurels in my line of conquests won ?
Must I, the belle of ballrooms, fade and pine,
    The maiden aunt to every coming son ?

It shall not be ! I will not be forgotten,
    Awake old charms to blooming life again !
With rouge and ribbons, corset, curls and cotton,
    I'll gird me for the Beauty's last campaign.

For hours she strove, with paint's unearthly flush,
    The ghost of buried loveliness to start ;
Then at the glass she madly dashed the brush
    And broke at once her mirror and her heart.

1843.

False Emblems.

THEY tell me friendships quickly formed
    Must early pass away,
Like dreams of bliss which cheer the night
    But fade with dawning day.
O be it so! for blissful dreams
    Around the heart will linger,
When life's realities are crushed
    By time's unsparing finger.

They tell me friendship's memories,
    How sweet soe'er they be,
Are transient as the purple glow
    Of sunset o'er the sea.
O be it so! for sunlight comes
    As gorgeous and as free,
As when it first in glory broke
    O'er earth's primeval sea.

Then be thy kindly thoughts of me
Like some sweet lingering dream,
Or like the light which fades at eve,
Once more with morn to beam.

1843.

Astrology.

" COULD I command the secret power
They say to stars is given,
Joy should attend thy every hour
Till lost in joys of heaven."

BREATHE not the wish ! diviner power
To faith's pure prayer is given,
Than clothed as old Chaldean deemed
The heraldry of heaven.

The kindly heart unfolds a page
　　More dear to sorrow's eye,
Than starry scroll where wise men read
　　The scripture of the sky.

And pity's tender voice doth fall
　　More sweetly on the ear,
Than harpings from the mystic Lyre
　　That leads each echoing sphere.

Oh then be thine the dearer charm
　　Of sympathy and love !
Be thine the nobler prayer that calls
　　Rich blessings from above.

And sigh not for the secret power
　　That dwells in worlds afar,
Be Faith thy sole astrologer,
　　And Hope thy guiding star.

For Hope, e'en like the orbs of night,
　　Can cheer life's loneliest way,
And Faith shall lead thee where the stars
　　Are lost in heavenly day.

1844.

Worship at Sea.

SOFTLY o'er the ocean
　　Faded sunset's ray,
Calmly on the waters
　　The idle vessel lay.
Hark ! as the soft winds rise
　　Mid twilight shadows dim,
They waft in music to the skies
　　The sailors' evening hymn !

Darkly brooding o'er them
　　Wave the storm-cloud's wings.
Forth the gathering tempest
　　His murky banner flings ;
Hark ! while the darkened sky
　　With thunder peals is riven,
Swells forth in voices calm and high
　　The sailors' prayer to heaven !

Brightly beams the morning
   O'er the vessel's track,
Every glancing ripple
   Gives its glory back;
Hark! round that altar lone
  Full, manly voices raise
On high, in solemn, thankful tone,
  The sailors' song of praise.

Thus on life's wide ocean
   Fearfully we sail,
Smiles or frowns above us
   Sunshine or the gale;
And thus in joy's fair morn
  Or sorrow's darkened sky,
Shall be our heart's free tribute borne
  With constant trust on high!

1844.

The Summer Cloud.

A FLEECY cloud as it rose on high
   In the path of the waking morn,
Like a phantom ship o'er the summer sky
   By the soft south wind was borne.

Gently and slowly that snow-white sail
   Swept on o'er the azure field,
Till it robed the sun with its silvery veil
   And shone like a golden shield.

An old blind man as he tottered by
   With years and their sorrows bowed.
With a smile turned upward his sightless eye
   And greeted that kindly cloud.

A laughing child on the grateful sight
   Looked forth from the shaded bowers,
And shouting for joy at the mellowed light,
   Ran gaily to gather flowers.

The pilgrim repented his solemn vow
　As he trod o'er the burning plain ;
But he blessed the cloud,as it cooled his brow,
　And his faith grew strong again.

The reaper stood in the golden grain
　Oppressed by the noon-day sun ;
The shadow fell, and he toiled again
　With a smile till his task was done.

The soldier failed in the fearful hour
　When the din of the strife grew loud,
But his arm was nerved as he felt the power
　That dwelt in that grateful cloud.

It followed the sun as his chariot rolled
　To the gates of the glowing West,
And gleaming there like a throne of gold
　It sank to its glorious rest.

Thus as the cloud of the summer day
　Glides on through the blazing dome,
Our bark of life on its mystic way
　Is borne to its destined home.

Oh thus to all hearts like the bountiful shade
   Be the joy of our sympathy given,
And thus with a glory that never shall fade
   Shall we rest in the mansions of Heaven.

1844.

To an old Oak in the Friends Cemetery. Salem

PRIDE of the ancient forests ! thy vast bough
   Hath waved of yore 'neath many a changing sky,
And in lone grandeur thou art lifting now
   Thine aged arms imploringly on high,
As though to call a blessing down from heaven
On loved ones to thy guardian shelter given.

Where are thy brethren ? when the sounding wood
   Through all its arches sent the warrior's cry,
Like bannered armies on the hills they stood
   While swept the baffled tempest idly by ;

And towering o'er them thou didst proudly stand,
Like some plumed chieftain of the forest land.

The spoilers came ; the " pomp of groves " is gone,
    The verdant crown around the mountain's brow
The forest robes across the valleys thrown,
    All with their native dust are blended now ;
They touched no leaf that crowned thy kingly head,
But laid beneath thy sheltering arms their dead.

Each year thy foliage falls upon the grave
    A golden robe to deck the halls of death,
And ceaselessly thy swaying branches wave
    In plaintive music to the soft wind's breath ;
Oh what could soothe the weeper's tortured brow,
Did friendship mourn as faithfully as thou ?

Mid spring's glad voices thou art sorrowing still,
    When happy birds to greet the morn are springing,
When every grove and vale and echoing hill
    With nature's joyous minstrelsy are ringing,
Thy leaves, like harp-strings tuned to notes of woe,
Sad requiems breathe o'er those who sleep below.

And still when winter's icy hand has thrown
  His shroud of snow above each lowly bed,
There wilt thou stand in solemn state alone,
  The white-robed guardian of the sleeping dead.
And the rude winds that hoarsely sweep along
Will wail through all thy boughs their dirge-like song.

Time hath not scathed thee ; o'er thy regal form
  The winter wind a hundred years hath passed,
And still for ages shalt thou brave the storm,
  Still shalt thou stand to battle with the blast ;
And long the stricken forms of earth shall come,
To seek beneath thy shade their final home.

Yet all thy power and pride shall pass away !
  Low in the dust thy lordly form shall bow,
Thy giant arms are weaker than decay,
  Though they can quell the whirlwind's fury now ;
While those whose requiem thou hast sung so long,
Shall rise and join in nature's funeral song.

1844.

On the death of Rev. E. G. Prescott.

RECTOR OF ST. JOHN'S CHURCH, SALEM, N. J.

HARK ! from the moaning sea
  A voice of sadness comes !
How fearfully its tone
  Falls on fond hearts and homes !
Our pastor and our friend
  Rests in his lonely grave,
And the winds their dirges blend
  With the wailing of the wave.

'Twas midnight on the deep
  When his spirit passed away,—
It soared from death's dark sleep
  To the light of endless day.
His dust by stranger hands
  To the sea's cold depths was given,—
But his soul by angel bands
  Was borne to its home in heaven.

Deep in the lonely main,
  They have laid his cherished form,
Unheeded sounds the dirge
  Of the sea-bird and the storm ;
But a mightier voice shall ring
  Through the ocean's solemn caves,
And that warning note shall bring
  The dead from their nameless graves.

In the heavenly courts that form
  In a glorious robe shall stand ;
Oh may we meet him there,—
  There in the brighter land !
Though we sleep beneath the wave,
  Or the sod where violets bloom,
May we find earth's lowliest grave
  But the portal of our home !

1844.

To a Comb.

WHY not meet for friendship's token,
  Guardian of that thoughtful brow ?
Fancies pure and dreams unspoken
  Cluster round thee now.

While the tresses thou hast parted
  Shade those calm and earnest eyes.
Thoughts of her, the gentle hearted
  In my soul shall rise.

Midst her locks thou oft hast nestled
  Fondly while she mused alone,
Heard'st thou not her inmost feelings
  Breathed in trembling tone ?

Knowest thou not the fairy dwellers
  In her fancy's secret home ?
Heard'st thou not their whispered voices ?
  Tell me, faithful comb.

Tell me all her wildest dreamings,
    Whisper all those tones again ;
Tell me, and I then will guard thee
    Fast in friendship's chain.

1846.

A Valentine to E. M.

SAY'ST thou 'tis a lover's duty
    By his glowing verse to prove
To thy heart the strength and beauty
    Of his plighted love ?

Think'st thou love can count its treasures
    In the common notes of song ?
Sound its depths with careless measures
    Now forgotten long ?

Breezes light the grasses bending
   Scatter tones around their path,
Sweeter than the storm-wind rending
   Forests in its wrath.

Sweetest music ever gusheth
   From the brook that brawls along,
While the brimming river rusheth
   Silently and strong.

Deeper thoughts and feelings dearer
   Than the lip of song can sing,
Still to thee, life's gentle cheerer,
   Evermore shall cling.

1847.

The Last Day of Nineteen.

TWENTY to-morrow! girlhood's hours are going,
   So calmly spent within thy cherished home,
And now, perchance, thy gentle eyes o'erflowing,
   Thou dreamest of the sterner days to come.

Thine hours of youth, those bright and careless hours
 Now crowding fast upon thy memory come,
Thy native hills, the trees, the friendly flowers
 And all the household voices of thy home.

And thou art sad! for many a cherished token
 Is fading on thy youth's receding shore,
And blessings breathed, and kindly warnings spoken,
 Shall fall like music on thine ear no more.

Weep on! I would not check a single tear,
 Whose brightness doth its purer fountain prove,
Yet would I whisper comfort to thine ear
 And to thy heart a word of earnest love.

Are there not visions of a nobler pleasure
 Than e'er hath filled thine eyes with grateful tears,
And blessings given in yet more bounteous measure,
 Than Heaven hath granted to thine earlier years?

Are there not higher thoughts of love and duty
 Of life's true work to do, its trials to bear,
That shed around thy path a holier beauty
 And summon from thy heart a purer prayer?

Are there not waiting hearts to cheer thy sadness,
  And one, of all, to whom thou wilt be given
To overflow his brimming cup of gladness
  And make it sparkle with the smile of heaven?

And though thy days of careless joy be ended,
  Though trials may come, and times of anxious care.
Dost thou not know on Whom thou hast depended,
  And will not mercy heed the voice of prayer?

Then welcome, welcome to my home and heart,
  Though scarce hath ebbed thy girlhood's joyous tide,
Thou still shall be to me life's " better part,"
  My heart's best friend, my counsellor, my guide.

And so when life's dark shadows round us lengthen,
  O may they shade a path of peace and love.
A love that time and trial shall only strengthen,
  A peace that speaks of endless rest above.

1848.

SUMMER.

Baptism of Tears.

It was a beautiful Sunday evening, the 10th after Trinity, 1852.

Dear L. lay in her own sweet room at rest for ever. At her feet was her little altar, with her holy books as she last had used them; her Bible and Prayer Book, her "Keble," "A Kempis," and "Holy living and dying." In the recess of the western window stood a table covered with "a fair linen cloth," and spread with vases of fresh white flowers, and a silver bowl filled with "pure water."

Through the half-closed shutter came the rays of the setting sun, and they brightened the flowers and glistened in the water, and then stretched across the room to encircle the calm and holy brow of "our darling."

All was ready for the holy service. The Priest came in and stood by the table. The little babe was brought, dressed in its white baptismal robes, and all our loved ones stood about it, and its poor heart-broken father knelt by the head of his sleeping wife.

For a moment all was still, while the Holy Spirit seemed to fill all our hearts with peace.

Then the prayers were said, the water sanctified, the vows made, our little treasure laid in the arms of Christ's minister, its precious mother's name given ; the water sprinkled on its little forehead " in the name of the Father, Son and Holy Ghost," the sign of the blessed Jesus made, while the little one smiled sweetly, and then it was given back into the arms of its mother's mother, a child of God, a member of Christ, and an inheritor of the kingdom of Heaven to be nursed for Jesus' sake.

We knelt again to yield our hearty thanks ; the Priest blessed us and again all was quiet, while the spirit of our sainted one seemed to come to us and bid us not to weep but to rejoice.   Afterwards when all had gone out and left our dear one again alone. a fresh breeze came in and strewed the holy water with white rose leaves.   It seemed as if an angel had scattered them there.

August 16th, 1852.   E. L. M. to BISHOP G. W. DOANE.

[In reply to the above letter the following was received from the Bishop.]

THE BAPTISM OF TEARS.

TENTH SUNDAY AFTER TRINITY, AUGUST 15, 1852.

" They that sow in tears, shall reap in joy."

THE lovely day had passed away,
Its stillness, on the landscape lay ;
A summer's sunset's lingering rays
Still kept the atmosphere, ablaze ;
When, gathered in a darkened room,
Where light just glimmered, through the gloom,
A sorrowing circle, silent sate :
Distressed, but not disconsolate.

But yesterday, and every grace,
That makes of home, a sacred place,
The comforts, and the charms of life,
That blend in Mother, and in Wife ;
All that the heart of man holds dear,
Was crowned and consecrated here.
Serene and beautiful, to-day,
Decked for the dead, our darling lay ;
Whose eye, whose soul, whose heart, had been
The charm of all this sacred scene ;

So calm, so sweet, our blessed dead,
We scarce could deem the spirit fled.
Like infant, tired, that sinks to rest,
At noon, upon its Mother's breast ;
Her score of summers scarcely done,
And yet, her crown of victory won.
It is her own, her charmèd room,
This ante-chamber of the tomb ;
Her Bible opens, at the day ;
The Book, that taught her how to pray,
Her Taylor, Kempis, Keble, lie
Just where she left them, all, to die.

In western window's deep retreat,
A table stands, in order meet,
With linen cloth, and roses white,
And crystal water, pure and bright.
The lingering beams of parting day,
Upon the trembling waters play ;
Then stretching through the glimmering gloom,
That fills the still, and sacred room.
Upon our dear one's forehead fall,
Like some celestial coronal ;
For sainted Mother, meet array,
To grace her babe's baptismal day.

Upon her fair and pulseless head,
His hand, the kneeling husband laid ;
The honored father bowed him low,
The mother's tears in silence flow.
From sisters, brothers, loved ones, friends,
The hushed and stifled sorrow blends ;
One heart, one voice, in faltering flow,
Pours the low litany of woe,
" Thou gavest, Thou hast taken, Lord,
We bless Thy Holy name and Word ! "

The surpliced Priest, comes gliding in ;
The wave is blessed that saves from sin,
It sparkles on an infant's brow,
The child of grace and glory, now,
The Mother's blessed name is given,
That one may serve for both, in Heaven ;
The cross is sealed, the pledge secured,
The heritage of Heaven, ensured ;
The Mother's arms, the treasure take,
With Jesu's mark, impressed, to nurse for Jesu's sake.

Scarce was the sacred service done,
And our dear dead one left alone,
When, whispering through the waving trees,
There came a balmy western breeze,

And strewed the rose-leaves, fair and white,
Upon the water, pure and bright.
As if some angel had been sent,
To certify the sacrament ;
And flowers of love and peace been given,
To strew our darling's path to Heaven ;
And way-marks left along the road.
To bring our baby, home to God.

RIVERSIDE. *August* 22. 1852.

At the Sea Shore.

TO E. L. M. ON HER BIRTHDAY.

WHILE I stand and muse beside the sea.
Murmuring restlessly forevermore,
Other tones are wafted back to me—
Voices from the far eternal shore.

Words of love I nevermore shall hear,
  Sounds of joy from earth forever gone,
Seem to float from some far distant sphere,
  Mingling with the ocean's ceaseless tone.

But the joyful paths o'er which I trod
  · When my darling walked the earth with me,
Ere her gentle soul went home to God,
  Seem like golden isles beyond the sea.

Lone upon the shore of life I stand,
  Wrecks of dearest hopes around me lie,
But life's fields of labor skirt the strand,
  Through them lies the pathway to the sky.

Then to thee, my own dear sister-wife,
  Turns my sorrowing heart with trust and love,
Clings to thee through all this weary strife,
  Looks to thee to guide my steps above.

Welcome, dearest, to my home and heart,
  None but thee could e'er be welcome there
In my memories thou alone hast part ;
  In my sorrows thou alone canst share.

Sharer of my griefs, dear comforter!
  Not for thee the light of " love's young dream,"
But the joy of duties done for her,
  Round thee like her gentle smile shall beam.

1854.

May.

SADLY through the blossoms
    Call we our sweet May;
May is all around you
    Buds and blossoms say.

Heed you not her foot-prints
    Wheresoe'er you pass;
Buttercups and violets
    Gleaming through the grass?

Feel you not her soft breath
  Through the leaflets play ?
Hear you not the blue-birds
  Singing, Welcome May ?

Decks she not her orchards
  With her robe of bloom ?
Fills she not the wild-wood
  With her rich perfume ?

Smiles not every daisy
  Through its dewy tear ?
Sings not every streamlet,
  May, sweet May, is here ?

Poureth not her sunshine
  From the fount of day ?
Why among the blossoms
  Call you then, Sweet May ?

Vainly do we call her,
  She is far away ;
Birds and brooks and blossoms
  Are not our Sweet May.

Blither than the blue-birds,
  Fairer than the flowers,
Gentler than the May breeze
  Whispering through the bowers,

Brighter than the sunshine,
  Merrier than May-Day,
Purer than the blossoms
  Was our darling May.

Thrice her little fingers
  Plucked the bright May flowers,
Round her thrice the blossoms
  Fell in fragrant showers,

Then e'er spring-buds opened,
  Took her, He who gave,
And he spreads the May flowers
  O'er her winter grave.

When no more the blossoms
  Wake from winter's tomb,
Still in heavenly gardens
  Our sweet May shall bloom.

1860.

Lizzie.

THE "Baptism of tears" was done.
The rites were closed, the Priest was gone ;
But still the blessed angel "sent
To certify the Sacrament"
Did vanish not amidst the gloom
That shadowed fast the " charméd room,"
But lingered there to watch and weep
Where mother lay in holy sleep ;
And hovering o'er the infant's head
He dried the tears that love had shed.
But the baptismal drops divine
He left in their own light to shine ;
And bending, waiting, listening there,
He heard the soft unspoken prayer,—
Oh sweet and blessed angel, stay !
And go not with the morn away ;

Always I had an angel near,
My sweet and precious mother dear ;
But she has gone beyond the sky.
No longer can she hear my cry ;
No more can she my footsteps lead.
Nor guard me in my hours of need.
Ah, who shall guide me on my way ?
Oh bright and gracious angel, stay !
The pitying spirit heard her cry
And sped not to his home on high ;
But yearning o'er that fleshly shrine
Now consecrate with holy sign,
Made it a living temple fair,
And dwelt a sacred presence there.
And morn by morn her opening eye
Beheld the angel standing by ;
And night by night her listening ear
Her angel's loving voice did hear.
He led her gently day by day
Through orphan childhood's dangerous way ;
He fed her soul with heavenly food,
Her mind with earth's most precious good ;
He led her forth in pastures green,
Beside "still waters," pure and clean,

That flowed within the cloistered bowers
Where bloomed the self-same snow-white flowers
His hand had strewn, with sunset's light,
On her baptismal water bright.
He drew upon her radiant face
The lines of rare celestial grace,
And poured into her soft brown eyes
A light from far beyond the skies ;
He taught her lovely lips a speech
That angel tongue alone could teach ;
And clothed with gifts and graces rare
A mortal form, for earth too fair—
Until a dower too rich was given,
Meet only for an heir of heaven.
And then—alas for hearts she left,
Of heavenly beauty twice bereft !
Her angel bore her hence away
Beyond the bounds of life's dark day,
To taste once more her mother's love
In waiting arms outstretched above.

1862.

AUTUMN.

Thirteen.

My Lulu in her teens to-day !
She flings her childish gauds away.
And puts the graver garments on
More meet for girlhood's early dawn ;
She looks with eager, wistful gaze.
Far on through life's enticing maze.
Full sure to quaff diviner joys
Than e'er she found in childhood's toys,
Through years to brighten more and more,
Till three and ten shall be three-score.
Oh be it so ! but if it be,
'Tis childhood's love and childhood's glee,
And childhood's simple faith and truth,
Will lend their charm to ripening youth,
And sweeten all the joys of life
For maiden, woman. mother, wife :
And childhood's closed but spotless page
Be read from teens to utmost age.

Nov. 17, 1877.

Sixteen.

My Fanny stands expectant on the verge
Which severs childhood's safe and sheltering bower
From the rude turmoil of life's outer world.
For sixteen years within that fold secure,
Nurtured in peace and guarded by strong love,
Her eyes have seen but shapes of loveliness,
Her tongue has uttered only songs of joy,
Her ears have heard but tones of tenderness,
Her hands, though busy, have plucked only flowers,
Her feet, though fast they tripped, have found no thorns.
But now, a change! her eager feet will press
Beyond the sheltering bound; her wistful eye
Will turn from old delights, and fondly scan
With a vague wonder all the tempting paths
That stretch far out across the fields of life.
Some path she soon must tread; oh could I choose
Which it shall be, and guard her safely there!
Too well I know that all the ways of life,

Though bright with flowers and cheered by purest joys,
But lead to heights of duty to be scaled.
Whatever path she treads, across it lies
The rugged " Mountain of the Holy Cross,"
Which she must climb before her journey ends.
Then go not yet, my child ! the way is long,
I cannot guide thee far ; some stronger love
May snatch thee from my arms, or I may faint
Upon the way, and leave thee there alone.
Still nestle here within the fold of home.

Nov. 18, 1878.

The First Diamond.

How bright her own first diamond shines,
    In maiden's eyes at sweet sixteen !
How fair the earth, how pure the sky,
    Reflected in its liquid sheen !

Across its steady shafts of light
    Her brief bright past no shadow throws :
And dancing in its crystal depths
    In shapes of joy, the future glows.

From this cold stone the light will shine
    Through smiles and tears through joy and care,
The rays that glow on golden curls
    Will gleam as bright on silver hair.

When steps shall fail and eyes grow dim,
    When youth is but a far-off dream,
This fount of light will flow as free,
    Its changeless rays as brightly gleam.

I give my child this shining stone,
    That one sure truth her heart may learn,—
From youth to age, through joy and pain,
    Through smiles and tears, my love shall burn.

CHRISTMAS, 1878.

A Clasp for a Christmas Cloak.

GRANDMOTHER dear, whose eyes have seen
   Now five and seventy Christmas morns,
And from whose path through all the years,
   The hands of love have plucked the thorns ;

What gift, in proof that our dear love
   Outweighs thy load of years, shall we,
Thy five and twenty children, bring,
   In worth and purpose meet for thee.

Not pearls—though fit for crown of queen :
   Not precious stones—though dazzling bright ;
The halo of thy rounded life
   Would quench for us the diamond's light.

Not books—though every page should burn
   With sages' lore and poets' fire ;
Thy days have nobler wisdom taught,
   Than sage's pen or poet's lyre.

62

Not costly web or curious vase—
   For all that minted gold could buy
Were poor beside thy garnered wealth—
   The treasures thou hast stored on high.

We give this cloak of warmest fur,
   To guard thy form from winter's blast ;
Thy mother-heart, with warmer clasp,
   Our love shall shield while life shall last.

CHRISTMAS, 1878.

My Wife's Crutches.

YE solemn, gaunt, ungainly crutches,
   That serve her frame such slippery tricks,
Were you within my lawful clutches,
   I'd fling you back in River Styx.

Ye grew beside the Boat of Charon,
  In murky fens of Stygian gloom,
Nor ever, like the rod of Aaron,
  Shall your grim spindles burst in bloom.

Your reeds were tuned for groans rheumatic,
  And croaking sighs from gouty man ;
Nor e'er shall thrill with tones ecstatic,
  As did the pipes of ancient Pan.

Avaunt you then, ye helpers dismal !
  Offend my eyes and ears no more ;
Go stalking back to realms abysmal,
  And guide the ghosts on Lethe's shore.

But see ! while yet my words upbraid them,
  Her crutches bud with blossoms fair,
And Patience, Love, and Faith have made them
  Than Aaron's Rod, more rich and rare.

And hark ! from out their hollows slender,
  No dismal groans or sighs proceed,—
But tones of joy more sweet and tender
  Than swelled from Pan's enchanted reed.

Then stay! your use her worth discloses,
    Your ghastly frames her worth transmutes,
From withered sticks, to stems of roses—
    From creaking reeds, to magic flutes.

JANUARY, 1879.

Fritz.

HAPPY, winsome little Fritz,
    Mamma's faithful crutchifer,
Softly round her knees he flits,
    Glad for e'en the touch of her.

Sober, solemn little Fritz,
    Only nine and yet so wise,
At her feet demure he sits,—
    Ears attent, and wondering eyes.

Jolly, merry little Fritz,
    Quick to catch the quips and jokes,
Laughing till his sides he splits,
    Giggling till he nearly chokes.

Gentle, loving little Fritz,
    Last of all the household line,
Frank and true, as well befits
    Polished gentleman of nine.

There, your birth-day portrait. Fritz,
    Painted by paternal touches,
Drains my rhymes, and strains my wits,—
    Go and carry mamma's crutches.

JANUARY 29. 1879.

The Acolyte.

BENEATH the shadows of the Porch,
    Within whose depths God's Altar stands,
The Acolyte uplifts the torch
    And bears it on with reverent hands.

He steps within the sacred rail,
   Where no unhallowed foot may tread ;
And fearless walks, where pride would quail,
   And sin would shrink with mortal dread.

He stands unshamed before the Ark—
   Upturns his pure, undazzled eyes ;
And, poising high the quivering spark,
   He light the fires of sacrifice.

Why stands this child in God's own place
   And feels no sense of human shame,
When even Moses hid his face,
   And shrank with awe from Horeb's flame ?

Such child-like faith might mount on High,
   And pass Heaven's Gate without appall ;
Like Samuel, answer, " Here am I,"
   If ev'n the Lord Himself should call.

Such eyes of innocence could gaze
   Unblenched on Altar fires above ;
And walk, like Shadrach, through the blaze,
   Unharmed, with Him whose name is Love.

FEBRUARY 24, 1879.

A Dirge for old St. Stephen's.

THE church that bears the martyr's name
    Beneath the axe and hammer falls ;
Its carven work the spoilers break,
    And ruthless hands destroy its walls.

Where swelled so long the organ tones,
    Now flow the voiceless waves of air,
And mute upon the soundless sod
    Now lies the tongue that called to prayer.

The saints whose faithful watch was kept, .
    With sandaled feet and solemn eyes,
Recoiling from the vandal touch,
    Have fled away to distant skies.

And Charity and Faith and Hope
    No more shall shed their blessings down
On heads before the altar bent,
    Beneath the pictured Cross and Crown.

Along those dim familiar aisles
　　No more the bridal train shall tread ;
No babe shall smile before the font,
　　No stricken mourner wail the dead.

Its form shall fade from human thought,
　　Its place be lost in coming days,
And weeds of toil and greed shall grow,
　　Where bloomed the flowers of prayer and praise.

Yet on this spot, in years to come,
　　Where haply other walls shall rise,
And other sounds of mart or home
　　Shall break the hush of evening skies,—

Some drops of balm will softly fall
　　To heal the wounds of souls in pain ;
Some lingering tones of love or hope
　　Shall stir the pulses once again.

Some life-worn man, whose wandering feet
　　Have lost the ways of faith and truth,
Will pause and bend his ear to catch
　　Faint echoes from his sinless youth.

Some woman, burdened with life's woes,
   Whose heart with bitter anguish swells,
Shall stop and lift her weary head
   To hear her far-off marriage bells.

No plough can raze the deep-drawn lines
   Where Christian soldiers waged their strife ;
And seeds long sown shall swell and bloom
   In soil where grew the Bread of life.

OCTOBER, 1879.